Quantitative Aptitude: Ratio & Proportionality

ZALGHI KHAN

Copyright © 2018 Zalghi Khan

All rights reserved.

ISBN: 13: 978-1717109767
ISBN:10: 1717109764

DEDICATION

This book is dedicated to all the extraordinary investment banks, such as JPMorgan Chase, Goldman Sachs, Deutsche Bank and the countless others that are underappreciated for the great service they provide. Thank you, God Bless. Special thanks goes also to Mai Le, the Goldman Sachs Associate for her inspirational role and work.

CONTENTS

Chapter 1	Introduction
Chapter 2	Pg 4 Historical Analysis of Ratio
Chapter 3	Pg 9 Ratios
Chapter 4	Pg 21 Ratios and Proportion (50 Major Questions with Explanations)
Chapter 5	Pg 80 Proportionality
Chapter 6	Pg 88 Direct & Inverse Proportionality Questions
Bibliography	Pg 102

INTRODUCTION

I have written this concise book in the genuine hope that my readers will comprehend one of the most profound mathematical concepts; as well as one of the most utilized in the world. Ratio and proportions are used throughout the vast and interrelated disciplines of finance, investment banking, accountancy and many others, in the sciences and engineering .

The concepts in this book are comprehensively explained, with a plethora of examples, questions and historical accounts. I believe those of us who have been educated in the disciplines of mathematics and sciences have a duty for the public understanding of such concepts, as well as for those who need a rigorous academic instruction in this topic.

The book is essential requirement for all students who are studying STEM[1] subjects at university. Additionally, those who are currently booked for examinations and assessments with one of the major banks need a thorough understanding of the ideas in this book. I trust that all will strongly benefit from this work.

---

[1] STEM is an acronym for Science, Technology, Engineering and Mathematics.

2 HISTORICAL ANALYSIS OF RATIO AND PROPORTION

The origin of the concept of ratios is difficult to ascertain with regards to its exact date and geographical location. The discussion of ratio in D.E Smith's History of Mathematics begins as the following:

"It is rather profitless to speculate as to the domain in which the concept of ratio first appeared. The idea that one tribe is twice as large as another and the idea that one leather strap is only half as long as another both involve the notion of ratio; both are such as would develop early in the history of the race, and yet one has to do with ratio of numbers and the other with the ratio of geometric magnitudes.

Indeed, when we come to the Greek writers we find Nicomachus including ratio in his arithmetic Eudoxus in his geometry, and Theon of Smyrna in his chapter on music."

The concept of a single human settlement being twice as large as another is so elementary that it most likely would have been understood in prehistoric society. Consequently, researching the origin of the concept of ratio is futile.

However, it is within the realm of possibility to trace the etymological precursor of the word "ratio" to the Ancient Greek λόγος (*logos*). Early translators rendered this into Latin as *ratio* ("reason"; as in the word "rational") (KNOWLEDGE ,2016).

As a corollary, we are unable to determine the circumstances which led to the ''discovery'' of this mathematical concept or the specific person[s] that gave rise to it. Writers of the medieval period utilized the term *proportio* i.e proportion to imply ratio; whilst the word proportionalitas (proportionality) was referenced to indicate the equality of ratios (Smith, 1923).

A theory of ratio and proportion was developed by the Pythagoreans as applied to number (Heath, 1908). Only rational numbers were part of the conceptual framework of numbers in the Pythagorean construct.

In Book V of the Elements[2], the most famous definition 5, takes centre stage; the following is a translation by Sir Thomas L Heath[3] :

Magnitudes are said to be in the same ratio, the first to the second and the third to the fourth, when, if any equimultiples whatever be taken of the first and third, and any equimultiples whatever of the second and fourth, the former equimultiples alike exceed, are alike equal to, or alike fall short of, the latter equimultiples respectively taken in corresponding order.

---

[2] The ***Elements*** is a renowned treatise of mathematics consisting of 13 books attributed to the ancient Greek mathematician Euclid in Alexandria Ptolemaic Egypt c. 300 BC. The work consists of postulates, definitions , propositions (**theorems** and constructions), and mathematical proofs of the propositions. The books cover plane and solid Euclidean geometry, elementary number theory , and incommensurable lines . *Elements* is the oldest extant large-scale deductive treatment of **mathematics.**

[3] **Sir Thomas Little Heath** (5 October 1861 – 16 March 1940) was a British civil servant, mathematician, classical scholar, historian of ancient Greek mathematics translator, and mountaineer. He was educated at Clifton College . Heath translated works of Euclid of Alexandria, Apollonius of Perga, Aristarchus of Samos and Archimedes of Syracuse into English.

Definition 5 is regarded as the definition of ratio by Euclid; however this is not true since Euclid does not mention an explicit definition of ratios (Euclid's theory of ratios, n.d.). Paradoxically this definition seems to be about the equality of ratios[4], not its meaning.

Euclid does not define ratios due to the fact that the Greek methodology of dealing with real numbers was primitive. They did not possess the decimal notation in any base; certainly not base 10. Their notions of limits and convergence were weak; as well as their notions of common fractions.

In conclusion, it is not possible to trace the origin of the concept of ratio to an exact period in time, person or place.

---

[4] Technical Point: mathematically, an equation stating the equality of ratios is defined as proportion.

ZALGHI KHAN

3 RATIO

Any rational analysis becomes impossible without the comprehension of meanings behind key concepts and terms. Prior to engaging in any intellectual exercise, it is a necessary prerequisite to have clarity of fundamental terms.

In contemporary parlance the following is a definition of Ratio from an authoritative source:

Ratio, Quotient of two values. The ratio of a to b can be written $a:b$ or as the fraction a/b. In either case, a is the antecedent and b the consequent. Ratios arise whenever comparisons are made. They are usually reduced to lowest terms for simplicity. Thus, a school with 1,000 students and 50 teachers has a student/teacher ratio of 20 to 1. The ratio of the width to the height of a rectangle is called an aspect ratio, an example of which is the golden ratio of classical architecture.

When two ratios are set equal to each other, the resulting equation is called a proportion

(Encyclopaedia Britannica, 2018)

Proportionality

Proportionality, In algebra, equality between two ratios. In the expression $a/b = c/d$, a and b are in the same proportion as c and d. A proportion is typically set up to solve a word problem in which one of its four quantities is unknown. It is solved by multiplying one numerator by the opposite denominator and equating the product to that of the other numerator and denominator. The term *proportionality* describes any relationship that is always in the same ratio. The number of apples in a crop, for example, is proportional to the number of trees in the orchard, the ratio of proportionality being the average number of apples per tree.

(Encyclopaedia Britannica, 2018)

Ratio and Proportion

The end of inquiry in mathematics is the acquisition of knowledge regarding unknown quantity[5]. This is achieved through the comparison of the unknown with other quantities equal to, less than or greater than the unknown. Equations and proportions play an important role in this.

When we make use of equations[6] (Oxford Dictionaries | English, 2018), we determine the quantity sought, by discovering

---

[5] In algebra, an **unknown** is represented by a letter. For example, in the equation $E = mc^2$, the letter m represents an **unknown** mass, and the letter E represents an **unknown** amount of energy.

[6] A statement that the values of two mathematical expressions are equal (indicated by the sign =).

its *equality* with some other quantity or quantities already known.

More frequently than not, there are occasions when we are to compare the unknown quantity with others which are *not equal* to it, but either greater or less. In this scenario we must apply a different method. In this methodology, the effort is applied to know *how much* one of the quantities is greater than the other, or *how many* times the one contains the other. The acquisition of this answer we discover a ratio of the two quantities. There are two types of ratios: Arithmetical and Geometrical both of which are applicable to arithmetic and geometry.

We need to understand ratios to understand proportion.

Difference between two quantities or set of quantities is defined as **Arithmetical ratio**. The actual quantities are called the terms of the ratio, the terms between which the ratio exists. Thus 2 is the arithmetical ratio of 5 to 3. This is expressed, by placing minus between the quantities thus, 5 - 3.

If both the terms of an arithmetical ratio be *multiplied* or *divided* by the same quantity, the *ratio* will, in effect, be multiplied or divided by that quantity. Thus if A-B=R, then multiply both sides by A, (Ax.4.)

$$\frac{a}{h} - \frac{b}{h} = \frac{r}{h}$$

If the terms of one arithmetical ratio be added to, or subtracted from, the corresponding terms of another, the ratio of their sum or difference will be equal to the sum or difference of the two ratios.

If a - b And d - h,
are the two ratios,
Then (a + d) - (b + h) = (a - b) + (d - h). For each = a + d - b - h.
And (a - d) - (b - h) = (a - b) - (d - h). For each = a - d - b + h.
    Thus the arith. ratio of 12 - 4 is 8
    And the arith. ratio of 5 - 2 is 3
The ratio of the sum of the terms 16 - 6 is 10, the sum of the ratios.
The ratio of the difference of the terms 6 - 2 is 4, the difference of the ratios.

The relationship between quantities expressed as the Quotient if one is divided by the other is defined as Geometric ratio.

As an example:

The ratio of 12 to 4, is 12/4 is 3. For this is the quotient of 12 divided by 4. In other words, it shows how often 4 is contained in 12.

Similarly, the ratio of any quantity to another may be expressed by dividing the former by the latter, or, which is the same thing, making the former the numerator of a fraction, and the latter the denominator.

Thus the ratio of a to b:

$$\frac{a}{b}$$

The ratio of d + h to b + c, is:

$$\frac{d+h}{b+c}$$

Another methodology of geometrical ratio expression is through use of double points one over the other between the two figures (quantities).

Here is an example : 24:8. This expresses the ratio of 24 to 8; the first quantity is the antecedent[7] and the latter the consequent[8] (Devlin, 2004).

The combined quantities are defined as a couplet.

If you have two out of these three the other one can be found: antecedent, consequent and the ratio

If a= the antecedent, c= the consequent, r= the ratio.

---

[7] An **antecedent** is the first half of a **hypothetical proposition**, whenever the if-clause precedes the then-clause. In some contexts the antecedent is called the ***protasis***.

[8] A **consequent** is the second half of a hypothetical **proposition**. In the standard form of such a proposition, it is the part that follows "then". In an **implication**, if P implies Q, then P is called the **antecedent** and Q is called the **consequent** .In some contexts the consequent is called the ***apodosis***.

By definition :

$$r = \frac{a}{c}$$

The ratio is equal to the antecedent divided by the consequent.

Multiplying by c, a = cr, that is, the antecedent is equal to the consequent multiplied into the ratio.

The consequent is equal to the antecedent divided by the ratio.

$$r, c = \frac{a}{r}$$

#1 In the event that two couplets have equality with regards to their antecedents in addition to their consequents, their ratios must be equal

#2 In the event that two couplets have equality of ratios; equality of antecedents; equality of consequents; and if the ratios are equal and if the consequents are equal, the antecedents are equal.

In the comparison of two quantities if there is an equality, the ratio is a unit or a ratio of equality. The ratio of 6.12:26 is a unit, for the quotient of any quantity divided by itself is 1.

In the situation the antecedent of a couplet is greater than the consequent, the ratio is greater than a unit. For if a dividend is greater than its divisor, the quotient is greater than a unit. Thus the ratio of 18:6 is 3. This is called a ratio of *greater inequality*.

Alternatively, if the antecedent is *less* than the consequent, the ratio is less than a unit, and is called a ratio of *less inequality*. Thus the ratio of 1:3, is less than a unit, because the dividend is less than the divisor.

Inverse ratio[9]

The change in the place with each other of the Antecedent and Consequent in a simple ratio, then the resultant ratio is called the **Inverse** of that ratio. As a corollary, the reciprocal ratio of **6 to 3**, is **1/6 :1/3.**

The reciprocal ratio is 1/a : 1/b or 1/a.b/1=b/a.

Compound Ratios

Two or more ratios multiplied term-wise i.e antecedent x antecedent and the consequent x consequent

Example:

Ratio of 6: 3 =2

Ratio of 12: 4= 3

Ratio compounded = 72: 12 =6

---

[9] Also referred to as Reciprocal Ratio

4 Ratios and Proportions (50 Major Questions with Explanations)

- 1. Find the fourth proportional to 2.4, 4.6 and 7.6?

A.) 14

B.) 14.657

C.) 15.56

D.) 14.56

Answer: Option 'D'

Formula= Fourth Proportional= (b. c)/a

A=2.4, B=4.6 and C=7.6

(4.6 *7.6)/2.4=14.56

- 2. Find the third proportional to 9 and 12?

A.) 9
B.) 108
C.) 16
D.) 9

Answer :Option 'C'
Formula= Third proportional= (b x b)/a
A=9 and B=12
(12 X 12)/9 = 144/9=16

- 3. Find the mean proportional between 49 & 81?

A.) 16
B.) 10
C.) 63
D.) 12

Answer: Option 'C'

Formula= √a×b
A=49 and B=81
√49×81=7×9=63

- 4. Find the fourth proportional to 0.2, 0.12 and 0.3?

A.) 0.13
B.) 0.18
C.) 0.8
D.) 0.15

Answer: Option 'B'
Formula= Fourth proportional =(b × c)/a
A=(0.12×0.3/0.2
0.036/0.2=0.18

- 5. If a:b=1:2 and b:c=3:4 find a:b:c?

A.) 3:8:6
B.) 3:12:6
C.) 3:6:8

D.) 1:2:4

Answer: Option 'C'
A:b=1:2, b:c=3:4
1:2
3:4
(a=1 × 3=3, b=2 × 3=6 and c=2 × 4=8)
(a=a × b, b=b × b and c= b × c)
a:b:c =3:6:8

- 6. If a:b=1:2, b:c=3:4 and c:d = 2:3 find a:b:c:d?

A.) 3:6:8:24
B.) 6:12:16:24
C.) 6:18:24:16
D.) 3:6:2:3

Answer: Option 'B'
A:b=1:2, b:c= 3:4, c:d=2:3
1:2

3:4
(a=1 ×3=3, b=2 × 3=6 and c=2 × 4 =8)
(a=a × b, b=b × b and c =b × c)
a:b:c =3:6:8
a:b:c= 3:6:8 and c:d =2:3
(Note: First a,b,c multiplication with c means 2 and last c means 8 multiplication with d means 3
a:b:c:d= 6:12:16:24

- 7. If a:b=2:3, b:c=4:5 and c:d=4:2 find a:b:c:d?

A.) 32:48:8:24

B.) 8:12:60:24

C.) 32:18:24:30

D.) 32:48:60:30

Answer: Option 'D'
a:b=2:3, b:c=4:5, c:d=4:2
2:3
4:5
(a=2 × 4 =8, b =37 × 4=12 and c=3 × 5 =15)
(a=a × b, b=b × b and c= b × c)
a:b:c= 8:12:15
a:b:c=8:12:15 and c:d=4:2

(Note: First a=8, b+12, c+15 multiplication with c means 4 and last c=15 multiplication with d means 2 a:b:c:d=32:48:60:30

- 8. If 2a=6b and 9b=7c, Find a:b:c?

A.) 42:18:14
B.) 3:9:7
C.) 9:27:22
D.) 21:7:9

Answer: Option 'D'
(2a=6b => a/b=6/2)
And (9b=7c => b/c=7/9)
=>a:b =6:2 and b:c =7:9
 a:b:c = 42:14:18=21:7:9

- 9. If 0.4:1.4 :: 1:4:x, then x=?

A.) 49
B.) 4.9
C.) 0.49
D.) 0.4

Answer: Option 'B'

0.4 × x=1.4 × 1.4
X=1.4 × 1.4/0.4=14/10 ×14/10 × 1/(4/10)
14/10 × 14/10 × 10/4
7/10 × 7=49/10=4.9

- 10. If x:y=5:3 then (8x-5y) : (8x+5y)=?

A.) 5:13

B.) 13:5

C.) 5:11

D.) 11:5

Answer: Option 'C'
x/y=5/3 (Given)
(8x-5y)/(8x+5y)
8(x/y)-5/8(x/y)+5
(on dividing Nr and Dr by y)
(8(5/3)-5)/(8(5/3)+5)
(40/3-5/1)/(40/3+5/1)

[(40-15)/3]/[(40+15)/3]
25/55=5/11
(8x-5y): (8x+5y)= 5:11

- 11. A fraction bears the same ratio to 1/27 as 3/7 does to 5/9. The fraction is?

A.) 7/45
B.) 1/35
C.) 45/7
D.) 5/21

Answer: Option 'B'
Let the fraction be x. Then,
 x:1/27=3/7: 5/9
X ×5/9=1/27 × 3/7
X × 5/9=1/9 × 1/7
X ×5/9=1/63
X × 5=9/63
5x=1/7=1/35

- 12. The ratio of two numbers is 3:4 and their sum is 28. The greater of the two numbers is?

A.) 8
B.) 12
C.) 14
D.) 16

Answer: Option 'D'
3:4
Total parts=7
=7 parts → 28(7 × 4=28)
=1 part → 4 (1× 4=4)
=The greater of the two number is =4

=4 parts → 16 (4 × 4=16)

- 13. The ratio of three numbers is 5:3:4 and their sum is 108. The second number of the three numbers is?

A.) 12
B.) 27
C.) 15
D.) 29

Answer: Option 'B'
5:3:4
Total Parts =12
12 Parts→ 108
1 Part →9
The second number of the three numbers is =3
3 parts → 27

- 14. Three numbers are in the ratio 3:5:7. The largest number value is 42. Find difference between Smallest & largest number is?

A.) 16
B.) 8
C.) 12
D.) 24

Answer: Option 'D'
=3:5:7
Total parts=15
=The largest number value is 42
=The largest number =7
=Then 7 parts → 42(7 × 6=42)
=smallest number = 3 & Largest number = 7
=Difference between smallest number & largest number is =7-3=4
= Then 4 parts → 24 (4 × 6=24)

- 15. If two numbers are in the ratio 2:3. If 10 is added to both of the numbers then the ratio becomes 3:4 then find the smallest number?

A.) 10
B.) 20
C.) 30
D.) 40

Answer: Option 'B'
2:3
2x+10 :3x+10=3:4
4[2x+10]=3 [3x +10]
8x+40 =9x +30
9x-8x=40-30
x=10
then smallest number is =2
2x=20
Short cut method:
a:b =2:3

c:d = 3:4
1. Cross multiplication with both ratios a × d ~ b ×c =2 *4 ~ 3 × 3= 8 ~9=1
2. If 10 is added both the number means 10 × 3=30 and 10 × 4=40,
Then 30 ~ 40 =10
  ⇨ 1→ 10
  ⇨ -> 20

- 16. If two numbers are in the ratio 2:3. If 10 is added to both of the numbers then the ratio becomes 5:7 then find the largest number?

A.) 30
B.) 10
C.) 60
D.) 40

Answer: Option 'D'

2:3
2x+10:3x+10=5:7
7[2x+10]=5[3x+10]
14x+70=15x+50
15x-14x=70-50
X=20
Then the first number is =2
2x=40
Short cut method:
a:b=2:3
c:d=5:7
1.Cross multiplication with both ratios
a×d~b×c=2×7~3×5=14~15=1
2. If 10 is added both the number means 10 × 5=50 and 10 × 7=70, then 50 ~ 70=20
=> 1→ 20
=> 2-→ 40 (Answer is =40)

- 17. If two numbers are in the ratio 5:3. If 10 is Reduced to both of the numbers then the ratio becomes 2:1 then find the smallest number?

A.) 10
B.) 20
C.) 30
D.) 40

Answer: Option 'C'

5:3
5x - 10 : 3x - 10 = 2 : 1
1[5x - 10] = 2[3x - 10]
5x - 10 = 6x - 20
6x - 5x = 20 - 10
x = 10
the small number is = 3
3x = 30 (Answer = 30)
Short cut method:
a:b = 5:3
c:d = 2:1
1.Cross multiplication with both ratios a × d ~ b × c = 5 × 1 ~ 3 × 2 = 5 ~ 6 = 1
2. If 10 is reduced both the number means 10 × 2 = 20 and 10 × 1 = 10,
Then 20 ~ 10 = 10
=> 1 ---> 10
=> 3 ---> 30 (Answer is = 30)

- 18. A mixture contains milk and water in the ratio 5:2. On adding 10 litres of water, the ratio of milk to water becomes 5:3. The quantity of milk in the original mixture is?

A.) 70

B.) 50
C.) 30
D.) 40

Answer: Option 'A'

milk: water = 5:2
5x : 2x + 10 = 5 : 3
3[5x] = 5[2x + 10]
15x = 10x + 50
15x - 10x = 50
x = 10
The quantity of milk in the original mixture is =
5 : 2 = 5 + 2 = 7
7x = 70
Short cut method:
milk: water = 5 :2
after adding 10 litres of water
milk: water = 5 :3
milk is same but water increase 10liters then the water ratio is increase 1 parts
1 part ---> 10 litres
The quantity of milk in the original mixture is =
5 : 2 = 5 + 2 = 7
7 parts ---> 70 litres (Answer is = 70)
Short cut method - 2 : for Only milk problems
milk : water
5 : 2
5 : 3

milk ratio same but water ratio 1 part incress per 10 litres
1 part of ratio ---> 10 litres
7 part of ratio ---> 70 litres

- 19. A mixture contains milk and water in the ratio 7:3. On adding 20 litres of water, the ratio of milk to water becomes 7:5. Total quantity of milk & water before adding water to it?

A.) 10
B.) 100
C.) 70
D.) 30

Answer: Option 'B'

milk: water = 7:3

7x : 3x + 20 = 7 : 5

5[7x] = 7[3x + 20]

35x = 21x + 140

35x - 21x = 140

14x = 140

x = 10

The quantity of milk in the original mixture is = 7 : 3 = 7 + 3 = 10

10x = 100

Short cut method:

milk: water = 7 : 3

after adding 20 litres of water

milk: water = 7 : 5

milk is same but water increase 20liters then the water ratio is increase 2 parts

1 part ---> 10 litres

The quantity of milk in the original mixture is =
7 : 3 = 7 + 3 = 10

10 parts ---> 100 litres (Answer is = 100)

Short cut method - 2 : for Only milk problems

milk : water

7 : 3

7 : 5

milk ratio same but water ratio 2 parts increase per 20 litres

2 part of ratio ---> 20 litres

1 part of ratio ---> 10 litres

10 part of ratio ---> 100 litres

- 20. A mixture contains milk and water in the ratio 3:2. On adding 10 liters of water, the ratio of milk to water becomes 2:3. Total quantity of milk & water before adding water to it?

A.) 10
B.) 50
C.) 20
D.) 40

Answer: Option 'C'

milk: water = 3:2

after adding 10 litres of water

milk: water = 2:3

Only water parts increase when mixture of water

milk: water = 3:2 = 2*(3:2) = 6:4

after adding 10 litres of water

milk: water = 2:3 = 3*(2:3) = 6:9

milk parts always same

Short cut method:

milk: water = 6 : 4

after adding 10 litres of water

milk: water = 6 : 9

milk is same but water increase 10liters then the water ratio is increase 5 parts

5 part ---> 10 litres

The quantity of milk in the original mixture is = 6 : 4 = 6 + 4 = 10

10 parts ---> 20 litres (Answer is = 20)

Short cut method - 2 : for Only milk problems

milk : water

6 : 4

6 : 9

milk ratio same but water ratio 5 parts increase per 10 litres

5 part of ratio ----> 10 litres

10 part of ratio ---> 20 litres

- 21. If Rs.900/- Rupees are divided among a,b and c in such a way that A's share 3 times that of B and B's share is 2 times that of C. The A's share is?

A.) Rs.600/-
B.) Rs.100/-
C.) Rs.200/-
D.) Rs.700/-

Answer: Option 'A'

A:B:C = 6:2:1

Total parts = 9

A's share is = 6 parts

9 ----> Rs.900/-

6 -----> Rs.600/-

- 22. If Rs.800/- Rupees are divided among a,b and c in such a way that A's share 4 times more than B, B's share is 3 times more than C. The C's share is?

A.) Rs.650/-
B.) Rs.600/-
C.) Rs.150/-
D.) Rs.50/

Answer: Option 'D'

A:B:C = 12:3:1

Total parts = 16

C's share is = 1 parts

16 -----> Rs.800/-

1 -----> Rs.50/- (Answer = Rs.50/-)

- 23. If Rs.540/- are divided among A,B and C in such a way that A's share is ½nd of B share and B's share is 1/3rd of C's share. The share of A is?

A.) Rs.80/-
B.) Rs.360/-
C.) Rs.60/-
D.) Rs.120/-

Answer: Option 'C'

A:B:C = 1:2:6

Total parts = 9

A's share is = 1 parts

9 -----> Rs.540/-

1 ----> Rs.60/-

- 24. If Rs.1440/- are divided among A,B and C so that A receives 1/3rd as much as B and B receives 1/4th as much as C. The amount B received is:

A.) Rs.90/-
B.) Rs.270/-
C.) Rs.1080/-
D.) Rs.27/-

Answer: Option 'B'

A:B:C = 1:3:12

Total parts = 16

B's share is = 3 parts

16 ----> 1440

1 ----> 90

3 ----> 270 (B's share is 270)

- 25. Rs.630/- distributed among A,B and C such that on decreasing their shares by RS.10, RS.5 and RS.15 respectively, The balance money would be divided among them in the ratio 3:4:5. Then, A's share is:?

A.) Rs.150/-
B.) Rs.200/-
C.) Rs.160/-
D.) Rs.255/-

Answer: Option 'C'

A:B:C = 3:4:5

Total parts = 12

A's share is = 3 parts

12 ----> Rs.600/-

3 ----> Rs.150/-

A's total = 150 + 10 = Rs.160/-

- 26. A bag contains 50p,Rs.1/- and Rs2/- coins in the ratio of 4:2:1 respectively. If the total money in the bag is Rs.60/-. Find the number of 50p coins in that bag?

A.) 60 coins
B.) 10 coins
C.) 20 coins
D.) 40 coins

Answer: Option 'D'

50paisa : Rs.1/- : Rs.2/- = 4 : 2 : 1 ---> coins ratio

= 2 : 2 : 2 ----> money ratio

Rs.2/- × 10 coins = Rs.20/-

Rs.1/- × 20 coins = Rs.20/-

50 Paisa × 40 coins = Rs.20/-

Then 50 paisa coins in that bag = 40 coins

- 27. A bag contains 5p, 10p and Rs20p coins in the ratio of 1:2:4 respectively. If the total money in the bag is Rs.84/-. Find the number of 10p coins in that bag?

A.) 160 coins
B.) 20 coins
C.) 128 coins
D.) None of these

Answer: Option 'A'

5paisa : 10paisa : 20paisa = 1 : 2 : 4 ---> coins ratio

= 5 : 20 : 80 ---> money ratio

5 paisa × 80 coins = Rs.4/-

10 paisa × 160 coins = Rs.16/-

Then 10 paisa coins in that bag = 160 coins

- 28. A bag contains 5p, 10p and Rs20p coins in the ratio of 4:2:1 respectively. If the total money in the bag is Rs.30/-. Find the number of 5p coins in that bag?

A.) 100 coins
B.) 150 coins
C.) 200 coins
D.) 250 coins

Answer: Option 'C'

5paisa : Rs.10/- : Rs.20/- = 4 : 2 : 1 ---> coins ratio

= 1 : 1 : 1 ---> money ratio

3 ---> 30

1 ---> Rs.10/- × 20 coins (1/- = 5paisa × 20 coins)

Rs.10/- = 200 coins(5paisa)

- 29. The ratio of first and second class train fares between two stations is 40:1 and that of the number of passengers travelling between these stations by first and second class is 1:20. If on a particular day Rs.2700/- be collected from the passengers travelling between these stations, then the amount collected from first class passengers is:?

A.) Rs.1800/-
B.) Rs.600/-
C.) Rs.900/-
D.) Rs.600/-

Answer: Option 'A'

First class: Second class = 40:1

Men's ratio = 1:20

Tickets ratio = 40 : 1

Men's ratio = = 1 : 20

Money ratio = = 40 : 20 ==> 2 : 1

3 parts -----> $.2700/-

1 part -------> $.900/-

2 parts -----> $.1800/-

- 30. If a dozen mirrors are fallen down. The ratio between broken and unbroken mirrors is:

A.) 2:3
B.) 3:4
C.) 5:7
D.) 5:4

Answer: Option 'C'

Dozen means = 12

12/(2+3) = 12/5 ==> wrong

12/(3+4) = 12/7 ==> wrong

12/(5+7) = 12/12 = 1 ==> Right (Answer = 5:7)

- 31. $.4800/- are divided among P,Q and R in such a way that the share of P is 5/11 of the combined share of Q and R. Thus, P gets:?

A.) $.300/-
B.) $.3300/-
C.) $.1800/-
D.) $.1500/-

Answer: Option 'D'

P/(Q+R) = 5/11

16 ----> $.4800/-

1 -----> $.300/-

P = 5 parts

5 -----> $.1500/- P = 300 × 5 = $.1500/-

- 32. $.4800/- are divided among P,Q and R in such a way that the share of P is 5/11 of the combined share of Q and R. The share of Q is 3/13 of the combined share of R and P. Thus, R gets:?

A.) $.300/-
B.) $.3300/-
C.) $.1500/-
D.) $.2400/-

Answer: Option 'D'

P/(Q+R) = 5/11

Q/(R+P) = 3/11

P:Q:R = 5:3:8

16 ------> Rs.4800/-

1 ------> Rs.300/-

R = 8 parts

R = Rs.300/- × 8 = Rs.2400/-

- 33. Two vessels of equal volumes contains milk and water mixed in the ratio 1:2,2:3. When These mixtures are mixed to form a new mixture, what is the ratio of milk and water?

A.) 11:19
B.) 19:11
C.) 2:5
D.) None of these

Answer: Option 'A'

1:2 , 2:3

1/3:2/3 , 2/5:3/5

= 1/3 + 2/5 : 2/3 + 3/5

= 11/15 : 19/15 = 11 : 19

- 34. A person spends one-third of the money with him on clothes, one-fifth of the remaining on food and one-fourth of the remaining on travel. Now, he is left with $. 100. How much did he have with him in the beginning?

A.) 200

B.) 250
C.) 300
D.) 450

Answer: Option 'B'

Initial amount be x.

Money spent on cloths = x/3.

Balance = x - (x/3) = 2x/3

Money on food, (1/5) x(2x/3) = 2x/15

Balance = (2x/3) - (2x/15) = 8x/15

Money spent on travel = (1/4) x (8x/15) = 2x/15

Balance = (8x/15) - (2x/15) =6x/15 = 2x/5

Given,2x/5 = 100

=> x = 250

Thus, the initial amount be $. 250

- 35. $. 770 have been divided among A, B and C such that A receives two-ninths of what B and C together receive. Then A's share is:

A.) $. 140
B.) $. 154
C.) $. 165
D.) $. 170

Answer: Option 'A'

Given, A = (2/9) (B+C)

=>(B+C) = 9A / 2

Given, A + B + C = 770

=>A + 9A/2 = 770

=> 11A = 770 x 2

=> A = 70 x 2 = 140

- 36. One-third of the contents of a container evaporated on the 1st day, three-fourths of the remaining evaporated on the second day. What part of the contents of the container is left at the end of the second day?

A.) One-fourth

B.) One-sixth

C.) One-half

D.) One-eighteenths

Answer: Option 'B'

**After first day, 2/3 rd of the contents remain
After second day 2/3 – (3/4) x (2/3) = 1/6 of the content remains**

- 37. If the ratio of boys to girls in a class is B and the ratio of girls to boys is G, then 3 (B + G) is

A.) Equal to 3

B.) Less than 3

C.) More than 3

D.) Less than one-third

Answer: Option 'C'

Boys = x, Girls = y x/ y = B and y/x = G 3(B+G) = 3(y/x + x/y = 3 (x2+y2)/xy > 3

- 38. Eight people are planning to share equally the cost of a rental car. If one person withdraws from the arrangement and the others share equally the entire cost of the car, then the share of each of the remaining persons increased by

A.) One-ninth
B.) One-eighth
C.) One-seventh
D.) Seven-eighths

Answer: Option 'C'

Given

When there are 8 people, the share of each person is 1/8

When there are 7 people, the share of each person is 1/7

Increase in the share of each person is

=>1/7 × 1/8

=> 1 / 56

Which if 1/7 of 1/8 of the original share of each person.

Share of each person = 1 / 7.

- 39.Trump purchased one dozen bangles. One day he slipped on the floor fell down. What cannot be the ratio of broken to unbroken bangles?

A.) 1 : 2
B.) 2 : 3
C.) 1 : 5
D.) 1 : 3

Answer: Option 'B'

There are totally 12 bangles,

Sum of the two numbers in the ratio should be a factor of 12,

Only 2:3 does not satisfy the criteria.

- 40. Two numbers are in the ratio of 1 :2. If 7 be added to both, their ratio changes to 3:5. The greater number is

A.) 20
B.) 24
C.) 28
D.) 32

Answer: Option 'C'

a/b = 1/2 and (a+7)/(b+7) = 3/5 => a= 14 and b = 28

- 41. In a mixture 60 litres, the ratio of milk and water 2 : 1. If this ratio is to be 1 : 2, then the quantity of water to be further added is:

A.) 20 litres

B.) 30 litres

C.) 40 litres

D.) 60 litres

Answer: Option 'D'

Quantity of milk = (60x2/3) litres = 40 litres

Quantity of water in it = (60- 40) litres = 20 litres.

New ratio = 1 : 2

Let quantity of water to be added further be x litres.

Then, milk : water = (40/ 20 + x)

Now, (40/20 + x = 80) => x = 60

Therefore Quantity of water to be added = 60 litres.

- 42. Seats for Mathematics, Physics and Biology in a school are in the ratio 5 : 7 : 8. There is a proposal to increase these seats by 40%, 50% and 75% respectively. What will be the ratio of increased seats?

A.) (2: 3: 4)
B.) (6: 7 : 8)
C.) (6 : 8: 9)
D.) (2 : 3 : 2)

Answer: Option 'A'

Originally, let the number of seats for Mathematics, Physics and Biology be 5x, 7x and 8x respectively.

Number of increased seats are (140% of 5x), (150% of 7x) and (175% of 8x).

=> (140 / 100 x 5x) , (150 /100 x 7x) and (175 / 100 x 8x)

=> 7x , 21x/2 and 14x .

Therefore, ratio of increased seats

= 7x : (21x/2) : 14x

= 14x : 21x: 28x

=2 : 3 : 4.

- 43. A sum of money is to be distributed among A, B, C, D in the proportion of 5 : 2 : 4 : 3. If C gets US Dollar 1000 more than D, what is B's share?

A.) $. 500

B.) $. 1500

C.) $. 2000

D.) $. 1200

Answer: Option 'C'

Let the shares of A, B, C and D be Rs. 5x, Rs. 2x, Rs. 4x and Rs. 3x respectively.

Then, 4x - 3x = 1000

=> x = 1000

QUANTITATIVE APTITUDE: RATIOS & PROPORTIONS

Therefore B's share

= Rs. 2x

= Rs. (2 x 1000)

= Rs. 2000.

- 44. Two numbers are respectively 20% and 50% more than a third number. The ratio of the two numbers is:

A.) 4:05

B.) 2:05

C.) 6:07

D.) 3:05

Answer: Option 'A'

Let the third number be x.

Then, first number = 120% of x

= 120x / 100

$= 6x/5$

Second number

$= 150\%$ of x

$= 150x/100$

$= 3x/2$

Therefore, Ratio of first two numbers

$= (6x/5 : 3x/2)$

$= 12x : 15x$

$= 4:5$

- 45. The ratio of the number of boys and girls at a party was 1:2 but when 2 boys and 2 girls left, the ratio became 1:3. Then the number of persons initially in the party was:

A.) 24

B.) 12

C.) 16

D.) 25

Answer: Option 'B'

12

- 46. The ratio of the number of boys and girls in a college is 7 : 8. If the percentage increase in the number of boys and girls be 20% and 10% respectively, what will be the new ratio?

A.) 21:22

B.) 8:09

C.) 17:18

D.) 8:11

Answer: Option 'A'

Originally, let the number of boys and girls in the college be 7x and 8x respectively.

Their increased number is (120% of 7x) and (110% of 8x)

=> (120 / 100 x 7x) and (110/ 100 x 8x)

=> 42x / 5 and 44x / 5

Therefore, The required ratio = (42 x/5 : 44x /5) = 42 : 44 = 21 : 22

- 47. Salaries of Ravi and Sumit are in the ratio 2 : 3. If the salary of each is increased by Rs. 4000, the new ratio becomes 40 : 57. What is Sumit's salary?

A.) Rs. 38,000

B.) Rs. 20,000

C.) Rs. 17,000

D.) Rs. 25,500

Answer: Option 'A'

Let the original salaries of Ravi and Sumit be Rs. 2x and Rs. 3x respectively.

Then, (2x+ 4000) / (3x + 4000) = 40 / 57

=> 57(2x + 4000) = 40(3x + 4000)

=> 6x = 68,000

=> 3x = 34,000

Sumit's present salary = (3x + 4000)

= Rs.(34000 + 4000)

= Rs. 38,000.

- 48. The sum of three numbers is 98. If the ratio of the first to second is 2 :3 and that of the second to the third is 5 : 8, then the second number is:

A.) 38

B.) 42

C.) 30

D.) 28

Answer: Option 'C'

Let the three numbers be A, B, C.

Then, A : B = 2 : 3 and

B : C = 5 : 8

Now, A : B = 2 : 3

=> A : B = (2 × 5) : (3 × 5) = 10 : 15

Now, B : C = 5 : 8

=> B : C = (5 × 3) : (8 × 3) = 15 : 24

Thus, A : B : C = 10 : 15 : 24

=> A = 10x, B = 15x, C = 24x

Given, A + B + C = 98

=> 10x + 15x + 24x = 98

=> 49x = 98

=> x = 2

Thus, Second number, B = 15x = 15 × 2 = 30

- 49. Two numbers are in the ratio 7:9. If 12 is subtracted from each of them, the ratio becomes 3:5. The product of the numbers is:

A.) 567

B.) 657

C.) 768

D.) 1575

Answer: Option 'A'

Given, ratio of two numbers = 7: 9 Let two numbers be 7x and 9x

If 12 is subtracted from each number, => (7x - 12) / (9x - 12) = 3/5 => 5(7x - 12) = 3(9x - 12) => 35x - 60 = 27x - 36 => 8x = 24 => x = 3

Product of the numbers = 7x × 9x = 7(3) × 9(3) = 21 × 27 = 567

- 50. Ratio between two numbers is 3 : 2 and their difference is 225, then the smaller number is:

A.) 90

B.) 450

C.) 270

D.) 480

Answer: Option 'B'

Given, the ratio of two numbers = 3 : 2

Let the two numbers be 3x and 2x

Given, their difference = 225

=> $3x - 2x = 225$

=> $x = 225$

Smaller number = $2x = 2 \times 225 = 450$

QUANTITATIVE APTITUDE: RATIOS & PROPORTIONS

Proportionality

Mathematically, proportionality is a quality that exists between two variables[10] when there is always a constant ratio between them. The constant is called the coefficient[11] of proportionality or proportionality constant.

---

[10] In **elementary mathematics**, a *variable* is a symbol, commonly an alphabetic character, that represents a number, called the *value* of the variable, which is either arbitrary, not fully specified, or unknown. In more advanced **mathematics**, a *variable* is a symbol that denotes a mathematical object, which could be a **number**, a **vector**, a **matrix**, or even a function.

[11] Constant value of the ratio of two proportional quantities x

Direct Proportionality

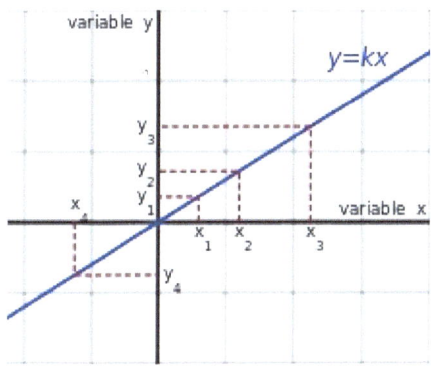

Figure 1 illustrates that variable y is in direct proportion to variable x. In the diagram, k represents the constant of proportionality. (Weisstein, 2018)

Direct Proportionality is a conceptually easy paradigm to comprehend- for example if we, double the volume of a solid we also double its mass.

If we are operating under the presupposition that when the variable x is 1, a corresponding variable y has the value 3; and when x is doubled so

---

and y; usually written y=kx, where k is the factor of proportionality that its value is 2, y also doubles so its value is 6. We could then postulate that when x is 3, y would have the value 9. This would indeed be the case if x is directly proportional to y.

Within the paradigm of y and x (two variables), x is directly proportional to y provided a non-zero constant (k) such that

$$Y = kx$$ [12]

---

[12] Standard Mathematical Denotation for Direct Proportionality: **y ∝ x**

$y \propto x$

$y = kx$ for a constant k

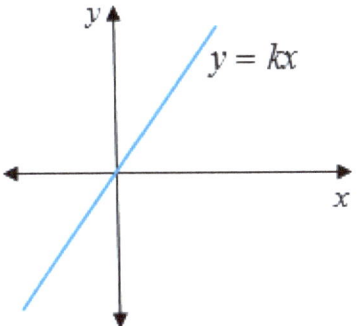

Figure 2 illustrates the equation of direct proportionality: y=kx

Hooke's Law States:

*" for relatively small deformations of an object, the displacement or size of the deformation is **directly proportional** to the deforming force or load."*

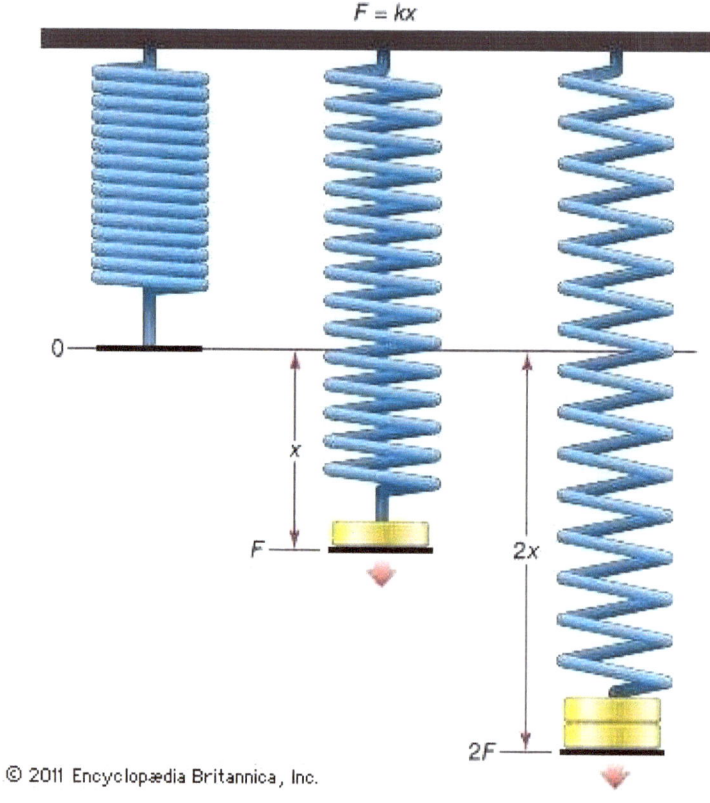

Figure 3 Hooke's law, $F = kx$, where the applied force F equals a constant k times the displacement or change in length x.

Inverse Proportionality

If two quantities, A, B, are **directly proportional**, then as you increase one the other also increases by a specific amount determined by a fixed number called the **constant of proportionality**. To denote this, we write $A \propto B$ (said "A is proportional to B"), from which you should be able to write down the equation $A = kB$, where k is the aforementioned constant of proportionality.

$$y \propto \frac{1}{x}$$

$$y = \frac{k}{x} \quad \text{for a constant } k$$

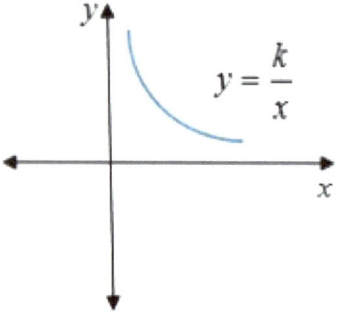

Fig 4

Two quantities y and x are said to be inversely proportional (or "in inverse proportion") if y is given by a constant multiple of $1/x$, i.e., $y = c/x$ for c a constant. This relationship is commonly written $y \propto x^{-1}$.

The graph of an inverse proportion is always a hyperbola[13], with asymptotes[14] at the x and y axes.

---

[13] Hyperbola is a conic section in which difference of distances of all the points from two fixed points (called foci) is constant. For two given points, the foci, a hyperbola is the locus of points such that the difference between the distances to each focus is constant.

[14] An asymptote is a line that a curve approaches, as it heads towards infinity. There are three types of asymptotes: horizontal, vertical and oblique. The crucial point is : the distance between the

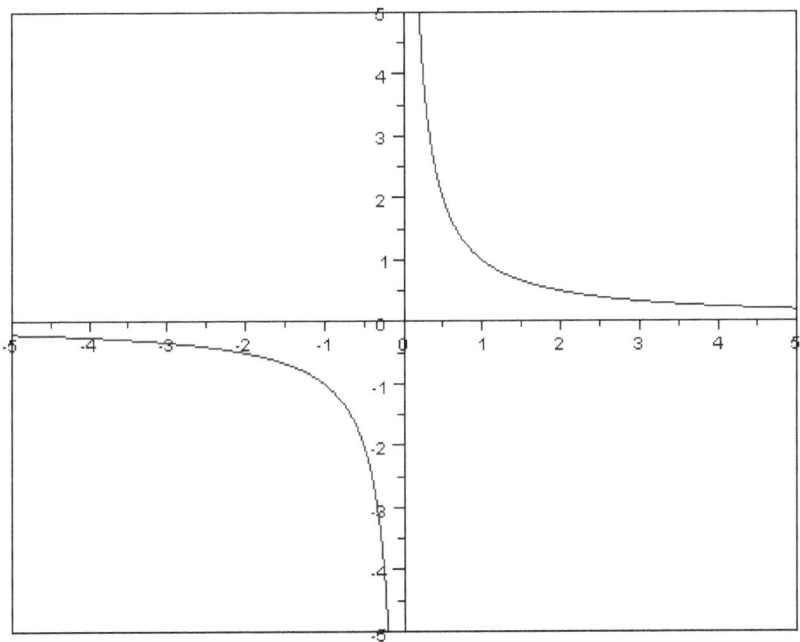

Figure 5 Inverse proportionality with a function of y=1/x

curve and the asymptote tends to zero as they head to infinity (or −infinity)

Direct and Inverse Proportionality Questions

01 – Directly proportional Q1

The force F on a mass is directly proportional ($\propto$) to the acceleration 'a' of the mass

When a = 350, F = 850

Find F when a = 140

02 – *Directly proportional Q2*

The resistance R (ohms) of a wire is directly $\propto$ to the length l (cm) of the wire

When l = 150, R = 750

Find R when l = 450

03 – *Directly proportional Q3*

The energy E (Joules) of a gas is directly $\propto$ to it's temperature T (Kelvin).

When T = 280, E = 40

Calculate T when E = 9

04 – *Directly proportional Q4*

The energy (E) released when matter is converted to energy is $\propto$ to mass of that object (m).

When E = 1.0×10^{16} Joules, m = 0.111 kg

Calculate the mass, in kg when E is 1.8 million Joules

Select the correct answer from the list below:

A: m = 2×10^{-11}

B: m = 2×10^{11}

C: m = 2×10^{-1}

05 – Inversely Proportional Q1

x is inversely proportional to y.

x = 5 when y = 12

Work out the value of y when x = 4

06 – Inversely Proportional Q2

The gravitational force F (Newtons) between two masses is inversely proportional to the square of the distance d between them.

When d = 8, F = 10

Calculate F when d = 10

07 – Inversely Proportional Q3

The time T (hours), required to build a brick wall is inversely proportional to the number of men M laying bricks.

When 6 men are laying bricks the wall takes 4 hours to build.

If it took ¾ hour to build the wall how many men would there be?

08 – Inversely Proportional Q4

The resistance R ohms of a wire is inversely proportional to the cross sectional area A cm^2 of the wire.

When A = 0.1, R = 180

Find A when R = 15

09 – Inversely proportional to square Q1

The time in minutes (T) for meals to be served at a busy restaurant is inversely proportional to the square of the number of waiters (W) working at that time.

It takes 20 minutes for meals to be served when 12 waiters are working.

Find an equation connecting T and W.

Select the correct answer from the list below:

A: T = 2880/W^2

B: T = W^2/2880

C: T = 2880W^2

10 – Inversely proportional to square Q2

2. The force F (Newtons) between two point charges is inversely proportional to the square of the distance d between them.

If $F = 2 \times 10^{-10}$ N when $d = 2 \times 10^{-3}$ m

Calculate F when $d = 4 \times 10^{-3}$ m giving your answer in standard form

Select the correct answer from the list below:

A: $F = 5 \times 10^{1}$

B: $F = 5 \times 10^{11}$

C: $F = 5 \times 10^{-11}$

11 – Proportional to the square Q1

The mass m of a liquid in a cylindrical container is proportional to the square of the radius r

When r = 7, m = 16

Find r when m = 9

Give your answer as a fraction in its simplest form.

12 – Directly proportional to cubed Q1

The mass of a solid sphere (M gm) is proportional to its radius (R cm) cubed.

When R = 6, M = 7560 gms

Find the value of M when R = 5

ANSWERS TO PROPORTIONALITY QUESTIONS

Q12 Answer = 4375

Q11 Answer = r 21/4

Q10 Answer F = 5×10^{-11}

Q9 Answer T = $2880/W^2$

Q8 Answer A = 12

Q7 Answer = 32 Men

Q6 Answer F = 6.4

Q5 Answer y = 15

Q4 Answer m = 2×10^{-11}

Q3 Answer T = 63

Q2 Answer R = 2250

Q1 Answer F=340

QUANTITATIVE APTITUDE: RATIOS & PROPORTIONS

Basic math symbols *(Rapidtables.com, 2018)*

Symbol	Symbol Name	Meaning / definition	Example
=	equals sign	equality	5 = 2+3 5 is equal to 2+3
≠	not equal sign	inequality	5 ≠ 4 5 is not equal to 4
≈	approximately equal	approximation	$sin(0.01) \approx 0.01$, $x \approx y$ means x is approximately equal to y
>	strict inequality	greater than	5 > 4 5 is greater than 4
<	strict inequality	less than	4 < 5 4 is less than 5
≥	inequality	greater than or equal to	$5 \geq 4$, $x \geq y$ means x is greater than or equal to y
≤	inequality	less than or equal to	$4 \leq 5$, $x \leq y$ means x is less than or equal to y
( )	parentheses	calculate expression inside first	2 × (3+5) = 16
[ ]	brackets	calculate expression inside first	[(1+2)×(1+5)] = 18
+	plus sign	addition	1 + 1 = 2
−	minus sign	subtraction	2 − 1 = 1
±	plus - minus	both plus and minus operations	3 ± 5 = 8 and -2
±	minus - plus	both minus and plus operations	3 ∓ 5 = -2 and 8

Symbol	Symbol Name	Meaning / definition	Example
*	asterisk	multiplication	$2 * 3 = 6$
×	times sign	multiplication	$2 \times 3 = 6$
·	multiplication dot	multiplication	$2 \cdot 3 = 6$
÷	division sign / obelus	division	$6 \div 2 = 3$
/	division slash	division	$6 / 2 = 3$
—	horizontal line	division / fraction	$\dfrac{6}{2} = 3$
mod	modulo	remainder calculation	7 mod 2 = 1
.	period	decimal point, decimal separator	2.56 = 2+56/100
a^b	power	exponent	$2^3 = 8$
a^b	caret	exponent	2 ^ 3 = 8
$\sqrt{a}$	square root	$\sqrt{a} \cdot \sqrt{a} = a$	$\sqrt{9} = \pm 3$
$\sqrt[3]{a}$	cube root	$\sqrt[3]{a} \cdot \sqrt[3]{a} \cdot \sqrt[3]{a} = a$	$\sqrt[3]{8} = 2$
$\sqrt[4]{a}$	fourth root	$\sqrt[4]{a} \cdot \sqrt[4]{a} \cdot \sqrt[4]{a} \cdot \sqrt[4]{a} = a$	$\sqrt[4]{16} = \pm 2$
$\sqrt[n]{a}$	n-th root (radical)		for $n=3$, $\sqrt[n]{8} = 2$
%	percent	1% = 1/100	10% × 30 = 3
‰	per-mille	1‰ = 1/1000 = 0.1%	10‰ × 30 = 0.3
ppm	per-million	1ppm = 1/1000000	10ppm × 30 = 0.0003

QUANTITATIVE APTITUDE: RATIOS & PROPORTIONS

Symbol	Symbol Name	Meaning / definition	Example
ppb	per-billion	1ppb = 1/1000000000	10ppb × 30 = 3×10^{-7}
ppt	per-trillion	1ppt = 10^{-12}	10ppt × 30 = 3×10^{-10}

Algebra symbols (Rapidtables.com, 2018)

Symbol	Symbol Name	Meaning / definition	Example
x	x variable	unknown value to find	when $2x = 4$, then $x = 2$
$\equiv$	equivalence	identical to	
$\triangleq$	equal by definition	equal by definition	
$:=$	equal by definition	equal by definition	
$\sim$	approximately equal	weak approximation	$11 \sim 10$
$\approx$	approximately equal	approximation	$sin(0.01) \approx 0.01$
$\propto$	proportional to	proportional to	$y \propto x$ when $y = kx$, k constant
∞	lemniscate	infinity symbol	
$\ll$	much less than	much less than	$1 \ll 1000000$
$\gg$	much greater than	much greater than	$1000000 \gg 1$

Symbol	Symbol Name	Meaning / definition	Example
( )	parentheses	calculate expression inside first	$2 * (3+5) = 16$
[ ]	brackets	calculate expression inside first	$[(1+2)*(1+5)] = 18$
{ }	braces	set	
$\lfloor x \rfloor$	floor brackets	rounds number to lower integer	$\lfloor 4.3 \rfloor = 4$
$\lceil x \rceil$	ceiling brackets	rounds number to upper integer	$\lceil 4.3 \rceil = 5$
$x!$	exclamation mark	factorial	$4! = 1*2*3*4 = 24$
$\mid x \mid$	single vertical bar	absolute value	$\mid -5 \mid = 5$
$f(x)$	function of x	maps values of x to f(x)	$f(x) = 3x+5$
$(f \circ g)$	function composition	$(f \circ g)(x) = f(g(x))$	$f(x)=3x, g(x)=x-1 \Rightarrow (f \circ g)(x)=3(x-1)$
(a,b)	open interval	$(a,b) = \{x \mid a < x < b\}$	$x \in (2,6)$
$[a,b]$	closed interval	$[a,b] = \{x \mid a \leq x \leq b\}$	$x \in [2,6]$
Δ	delta	change / difference	$\Delta t = t_1 - t_0$

QUANTITATIVE APTITUDE: RATIOS & PROPORTIONS

Symbol	Symbol Name	Meaning / definition	Example
Δ	discriminant	$\Delta = b^2 - 4ac$	
Σ	sigma	summation - sum of all values in range of series	$\sum x_i = x_1 + x_2 + \ldots + x_n$
ΣΣ	sigma	double summation	$\sum_{j=1}^{2} \sum_{i=1}^{8} x_{i,j} = \sum_{i=1}^{8} x_{i,1} + \sum_{i=1}^{8} x_{i,2}$
∏	capital pi	product - product of all values in range of series	$\prod x_i = x_1 \cdot x_2 \cdot \ldots \cdot x_n$
e	e constant / Euler's number	$e = 2.718281828\ldots$	$e = \lim (1+1/x)^x$, $x \to \infty$
γ	Euler-Mascheroni constant	$\gamma = 0.5772156649\ldots$	
φ	golden ratio	golden ratio constant	
π	pi constant	$\pi = 3.141592654\ldots$ is the ratio between the circumference and diameter of a circle	$c = \pi \cdot d = 2 \cdot \pi \cdot r$

BIBLIOGRAPHY

ENCYCLOPEDIA BRITANNICA. (2018). *RATIO | MATHEMATICS*. [ONLINE] AVAILABLE AT: HTTPS://WWW.BRITANNICA.COM/SCIENCE/RATIO [ACCESSED 17 APR. 2018].

ENCYCLOPEDIA BRITANNICA. (2018). *PROPORTIONALITY | MATHEMATICS*. [ONLINE] AVAILABLE AT: HTTPS://WWW.BRITANNICA.COM/SCIENCE/PROPORTIONALITY [ACCESSED 17 APR. 2018].

MATHEMATICSDICTIONARY.COM. (2018). *MATH DICTIONARY*. [ONLINE] AVAILABLE AT: HTTP://WWW.MATHEMATICSDICTIONARY.COM/MATH-VOCABULARY.HTM [ACCESSED 17 APR. 2018].

WEISSTEIN, E. (2018). DIRECTLY PROPORTIONAL. [ONLINE] HTTP://MATHWORLD.WOLFRAM.COM. AVAILABLE AT: HTTP://MATHWORLD.WOLFRAM.COM/DIRECTLYPROPORTIONAL.HTML [ACCESSED 18 APR. 2018].

KNOWLEDGE, S. (2016). PENNY CYCLOPEDIA OF THE SOCIETY FOR THE DIFFUSION OF USEFUL KNOWLEDGE,. [S.L.]: FORGOTTEN BOOKS, P.307.

Themathpage.com. (2018). *Proportionality -- A complete course in arithmetic.* [online] Available at: http://www.themathpage.com/arith/proportionality.htm [Accessed 18 Apr. 2018].

Oxford Dictionaries | English. (2018). *proportion | Definition of proportion in English by Oxford Dictionaries.* [online] Available at: https://en.oxforddictionaries.com/definition/proportion [Accessed 18 Apr. 2018].

Smith, D. (1923). *History of mathematics.* Boston (MA), [etc.]: Ginn, p.478.

Euclid Euclid (2012). *The Thirteen Books of the Elements, Vol. 1.* Dover Publications.

heath, S. (1908). *The thirteen books of Euclid's Elements*. 2nd ed. Cambridge Univ. Press. pp. 112ff., p.112.

EUCLID'S THEORY OF RATIOS. (N.D.). .

OXFORD DICTIONARIES | ENGLISH. (2018). *EQUATION | DEFINITION OF EQUATION IN ENGLISH BY OXFORD DICTIONARIES*. [ONLINE] AVAILABLE AT: HTTPS://EN.OXFORDDICTIONARIES.COM/DEFINITION/EQUATION [ACCESSED 19 APR. 2018].

SETS, FUNCTIONS AND LOGIC - AN INTRODUCTION TO ABSTRACT MATHEMATICS, KEITH DEVLIN, CHAPMAN & HALL/CRC MATHEMATICS, 3RD ED., 2004

DEVLIN, K. (2004). *SETS, FUNCTIONS, AND LOGIC*. 3RD ED. BOCA RATON, FLA.: CHAPMAN & HALL/CRC.

https://www.onlinemathlearning.com/ratio-math-problems.html

https://artofproblemsolving.com/wiki/index.php?title=Proportion#See_also

https://mathsmadeeasy.co.uk/tests/direct-and-inverse-proportion-revision/question-04-23/

http://www.msgarciamath.com/proportion/proportional-relationships/

https://www.theatlantic.com/international/archive/2013/03/russia-is-keeping-its-elites-on-a-shorter-leash/273965/

https://www.learner.org/workshops/algebra/workshop7/index2.html

Yates, P. (2015). Y = mx + c. [image] Available at: https://eic.rsc.org/maths/a-sense-of-proportion/2000158.article [Accessed 23 Apr. 2018].

RAPIDTABLES.COM. (2018). MATHEMATICAL SYMBOLS LIST (+,-,X,/,=,<,>,...). [ONLINE] AVAILABLE AT: HTTPS://WWW.RAPIDTABLES.COM/MATH/SYMBOLS/BASIC_MATH_SYMBOLS.HTML [ACCESSED 23 APR. 2018].

ABOUT THE AUTHOR

Zalghi Khan is a mathematician, author and student. He has authored numerous academic papers on investment banking and a book on power strategies titled "Profile of Power: Vladimir Putin". In his spare time he helps raise awareness of the importance of Cancer Research, practices martial arts and Zen Meditation.

www.ingramcontent.com/pod-product-compliance
Lightning Source LLC
Chambersburg PA
CBHW040219220526
45473CB00001B/49